YOUR KNOWLEDGE HAS VALUE

- We will publish your bachelor's and master's thesis, essays and papers

- Your own eBook and book - sold worldwide in all relevant shops

- Earn money with each sale

Upload your text at www.GRIN.com and publish for free

Bibliographic information published by the German National Library:

The German National Library lists this publication in the National Bibliography; detailed bibliographic data are available on the Internet at http://dnb.dnb.de .

Imprint:

Copyright © 2016 GRIN Verlag, Open Publishing GmbH
Print and binding: Books on Demand GmbH, Norderstedt Germany
ISBN: 9783668404656

This book at GRIN:

http://www.grin.com/en/e-book/354281/production-of-bioethanol-from-agricultural-waste-biomass

Kevin Schreier

Production of bioethanol from agricultural waste biomass

GRIN Publishing

Bioenergy – Production of bioethanol from agricultural waste biomass

Kevin Schreier

(Masterstudium Energietechnik, 9. Semester)

ABSTRACT

The need of bioenergy due to the world's increasing population and the limited fossil energy resources, which by combustion are damaging the environment, leads to the demand of renewable energy resources. Waste biomass, especially from agriculture, is an underestimated, but attractive alternative to food-crops for the sustainable production of ethanol from biomass and able to replace petroleum-based fuels. The conversion-technology of agricultural waste biomass to bioethanol is already at the demonstration-stage. Compared to first generation bioethanol, the second generation fuel requires a more complex preparation-step of the feedstock. Through its wide range of application is bioethanol already applied worldwide and being produced by waste biomass its future looks promising.

Keywords: waste biomass, bioethanol, conversion technologies, biofuel, renewable energy sources

Content

Content... 2

Introduction: the need of Bionenergy - an opportunity for waste biomass 3

Agricultural waste biomass – an investigation 4

    Biomass Composition 4

Bioethanol – a desired biofuel and a product of waste biomass 5

    Bioethanol 5

    Conversion Technologies... 6

    Conversion of 1st generation bioethanol... 7

    Conversion of 2nd generation bioethanol 7

2nd generation technologies have the ambition to produce sustainable and affordable biofuels from resources which do not compete with food production. 8

    Application fields 8

Image-References.. 9

Introduction: the need of Bionenergy - an opportunity for waste biomass[1]

The world's increasing population enhances the need for more energy to be produced.

As a matter of fact, fossil energy resources are limited and their use is hugely damaging our environment. Consequently, there is an increasing demand for renewable energy sources, with less impact on the environment. Renewable energy sources use energy sources that are continually replenished by nature – the sun, the wind, water, the Earth's heat, and plants. Customary sources of renewable energy are solar, wind, geothermal, hydro, and some forms of biomass.[2] Thus Biomass represents a possibility to approach this problem. Biomass is organic material like trees, agricultural residues, animal wastes, grass, aquatic plants, and municipal waste. Through photosynthesis Biomass stores the energy from the sun. Bioenergy technology extracts the energy stored in biofuels through direct combustion or by converting the fuel into charcoal, liquid, or gas.

Two of the main methods to save the environment are simply saving energy by using more efficient technologies and secondly saving greenhouse gas emissions. The conversion from biomass waste to energy reduces greenhouse gas emissions twofold: Heat and electrical energy is produced, which diminishes the dependence on power plants based on fossil fuels. Further the greenhouse gas emissions are significantly reduced by avoiding methane emissions into the air from landfills.

Thus, there is a raising interest in biomass as part of the renewable energy resources in order to produce bioenergy. The use of biomass such as food-crops is connected with the problem of higher food prices. Using waste biomass is an attractive alternative. There is a high potential in waste biomass, for there is a lot of waste biomass currently just used as disposal in landfill or as animal food. Large quantities of crop residues are produced annually worldwide, and are vastly underutilized. Waste disposal becomes more and more expensive and in many places there is even not enough land.

There are many reasons for using waste biomass in order to produce bioethanol – a needed motor-fuel and attractive energy storage.

---

[1] A. Demirbas, Competitive liquid biofuels from biomass, Appl Energy, 88 (2011), pp. 17–28
[2] Dollars From Sense: The Economic Benefits of Renewable Energy.
http://www.nrel.gov/docs/legosti/fy97/20505.pdf

Agricultural waste biomass – an investigation[3]

In general the term biomass covers all materials deriving from plants that use sunlight to grow. Biomass can be used in many different ways such as food, construction material, medicine, fertilizer, fibre or energy.

In the context of energy-production, biomass is often applied to plant-based materials, but the term biomass can also refer to animal and vegetable-derived materials and waste.

Biomass resources are numerous such as plants, wood plants, food crops, energy crops and herbaceous plants etc.

Agricultural waste biomass are left-overs from agricultural processes and industry. Agricultural waste biomass covers grasses and flowers (like arundo, bamboo, bana, cane, brassica, miscanthus), straws (barley, rice, wheat, sunflower, oat, bean) and other residues (like fruits, shells, pits, grains, seeds, cobs, puls).

It is possible to divide agricultural residues into field-based residues and process-based residues. Biomass, generated in the agricultural field or farm is defined as field-based. Process-based residues are biomass, which is generated in the process of agricultural products (e.g. rice husk, bagasse, maize cob, peanuts shell).

Fig. 1: Bagasse as an example for process-based residues. [1].

Process-based residues are usually a lot more available than field-based waste biomass. Thus it is used as an energy source for the agricultural industry.

Biomass Composition

---

[3] Bisma Malik et al., Biomass Pellet Technology: A Green Approach for Sustainable Development, p.405ff.

Biomass uses sunlight to grow. As a result of the photosynthesis plants are able to store solar energy in the form of chemical energy while they grow. The components of biomass include a lot of different, complex compounds like cellulose, hemicelluloses, lignin, proteins, lipids, sugar, water and other compounds.

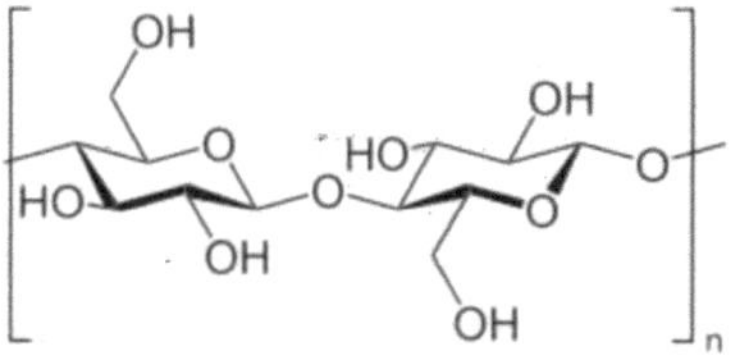

Fig. 2: Cellulose, a polysaccharide composed of two D-glucose linked together, is a major compound of waste biomass [2].

Cellulose, hemicelluloses and lignin are the main compounds of biomass. The amount and distribution of the compounds in the biomass waste varies significantly from species to species. In table 1 you can see typical distributions of these major compounds for three different type of biomass.

Table 1: Typical compounds of some agricultural biomass

Biomass	Cel-luloses	Hemicel-luloses	Lignin
Orchard Straw	32.0	40.0	4.7
Rice straw	34.0	27.2	14.2
Birch -wood	40.0	25.7	15.7

Bioethanol – a desired biofuel and a product of waste biomass

Bioethanol

When Biomass is processed and upgraded, it becomes a biofuel, which may be used instead of petroleum-based liquid fuels.

Biodiesel, renewable diesel and bioethanol are the main biofuels.[4] In this paper bioethanol is focused. Bioethanol is the most popular type of bioalcohol produced by using enzymes and microorganisms through fermentation of biomass. Among the second generation biofuels it plays a principle role. In general, bioethanol is made from lignocellulosic biomass, woody crops and agricultural residues. Thus the main advantage of bioethanol is that it can be sustainably produced by using biomass consisting of the residual non-food parts of crops. By using agricultural waste biomass it is not competing with the food-industry. However the chemical structure of bioethanol is just like that of ethanol, which is depicted in figure 3.

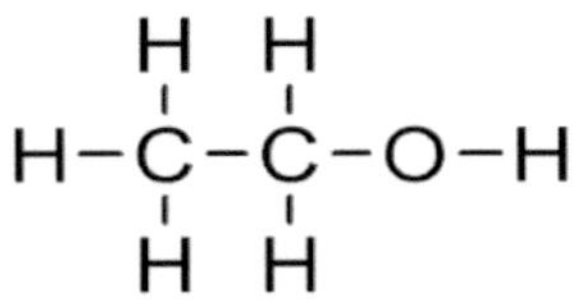

$$H-\underset{\underset{\displaystyle H}{|}}{\overset{\overset{\displaystyle H}{|}}{C}}-\underset{\underset{\displaystyle H}{|}}{\overset{\overset{\displaystyle H}{|}}{C}}-O-H$$

Fig. 3: The chemical structure of Ethanol: C_2H_5OH [3].

Conversion Technologies

Bioethanol is a natural product, which is generally produced by fermentation sugar-containing plants. There are several different conversion technologies that are adapted to the different physical nature and chemical composition of the feedstock.[5] Thus the conversion technology depends on the composition of the applied raw-material. If the raw-material mainly contains sugar and starch, the manufactured bioethanol is a so called first generation biofuel. Second-generation bioethanol derives from lignocellulosic biomass[6]. Whether first or second generation fuels, bioethanol is generally produced in four steps:

---

[4] M. Patel, A. Kumar, Production of renewable diesel through the hydroprocessing of lignocellulosic biomass-derived bio-oil: A review, p.1294

[5] A. Bauen, G. Berndes et al., Bioenergy – a sustainable and reliable energy source, a review of status and prospects, p.8

[6] G. Sivakumar, D. R. Vail, J. Xu et al, Bioethanol and biodiesel: Alternative liquid fuels for future generations

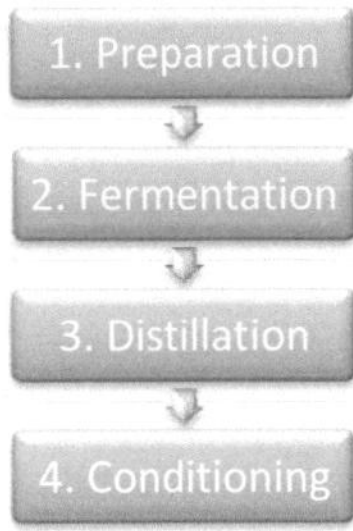

Fig. 4: The general conversion path from biomass to bioethanol

conversion technologies commonly begin with a preparation as their first step. Thereupon follows the fermentation process conducted by microorganism. Subsequently, the distillation takes place. The fourth and last step is the conditioning of the resulting bioethanol.

Conversion of 1st generation bioethanol[7]

Conversion technologies of this bioethanol based on sugar and starch containing feedstock are well-developed and already used in an industrial scale. The preparation of this conversion technology needs to be distinguished from the preparation of the conversion technology of 2nd generation bioethanol: the crop is ground in a mill, heated and water with enzymes for saccharifying is added. Step two to four are the same again. To get the fermentation started yeast is added to the mash. By fermentation the sugar is converted into ethanol and CO_2. This biological fermentation process is technically mature and commercially available. The third step, the distillation is to separate and to concentrate the resulting ethanol from the water. The last step is a conditioning-step. It is usually dehydration of the bioethanol, but highly depending on its final application.

Conversion of 2nd generation bioethanol[8]

2nd generation biofuels comprise a wide range of novel biofuels based on new feedstocks like novel starch, oil or sugar crops such as Jatropha, cassava or Miscanthus. Further it

---

[7] A. Bauen, G. Berndes et al., Bioenergy – a sustainable and reliable energy source, a review of status and prospects, p.2 and p.33

[8] A. Bauen, G. Berndes et al., Bioenergy – a sustainable and reliable energy source, a review of status and prospects, p.33ff.

encompasses numerous conventional and novel biofuels, for example ethanol, butane and syndiesel.

2nd generation bioethanol is produced from lignocellulosic materials. This is any organic matter which contains a combination of lignin, cellulose and hemicelluloses, which includes agricultural wastes (e.g. straw), energy crops (e.g. Miscanthus, poplar), forestry products and wastes and parts of municipal solid waste. The conversion technologies for 2nd generation bioethanol are still at the demonstration stage.

2nd generation technologies have the ambition to produce sustainable and affordable biofuels from resources which do not compete with food production.

In difference to the conversion technology of 1st generation bioethanol, 2nd generation bioethanol is produced by first cracking the cellulose and hemicellulose into sugars, which then can be fermented like it is done in the process of 1st generation bioethanol. There is a more advanced preparation needed than it is in the production of 1st generation bioethanol, for Lignocellulosic materials are more complex to break down than starch is.

Application fields

There are many different fields, where Bioethanol can be used. As bioethanol has the same structural properties, its application fields comprises all those of ordinary ethanol. Consequently, the principal application field of bioethanol is the fuel sector.

In the European Union a Fuel-Quality-Directive has been introduced to establish bioethanol as a fuel. Blended with petrol it is sold as an additive called "E5" or "E10". E10 means that 10 -vol. % of bioethanol are blended in petrol. Over 80% of the world's ethanol production is used in the fuel sector.[9] Bioethanol is an alternative fuel, able to replace customary ethanol.

Moreover bioethanol can be applied in food industry. As ethanol is both, hydro- and lipophilic, it may be used as organic solvent. Further it is a disinfectant and a cleaning agent. Lastly, ethanol is a useful basic material of the chemical industry, especially for the synthesis of other products.

---

[9] Source: http://www.cropenergies.com/en/Bioethanol/-Verwendung/, 21th June 2016

Image-References

Fig.[1] http://www.hxcorp.com.vn/news/1346-electricity-from-bagasse-every-year-can-save-500-mw-power.html, 31st of May 2016

Fig.[2] http://www.chemgapedia.de/vsengine/-media/width/296/height/133/vsc/de/ch/9/mac/copolymere/polymeranalog/cellulose.svg.jpg

Fig.[3] http://www.lte.lu/chimie/9ST_e/cours/ 00teilch/form/form.htma, 23rd of June 2016

Fig.[4] is of my own making. 21st of June 201